Bibliographic information published by the German National Library:

The German National Library lists this publication in the National Bibliography; detailed bibliographic data are available on the Internet at http://dnb.dnb.de .

Imprint:

Print and binding: Books on Demand GmbH, Norderstedt Germany
ISBN: 9783668590069

This book at GRIN:

http://www.grin.com/en/e-book/378731/modifications-of-graphene-oxide-based-composites-biomaterial-promote-human

Y. Umul Hanim

Modifications of Graphene Oxide-based composites biomaterial promote human Wharton's Jelly derived Mesenchymal Stem Cells. Functionalization and characterization

GRIN Publishing

Contents

Modifications of Graphene Oxide-based composites biomaterial promote human Wharton's Jelly derived Mesenchymal Stem Cells: functionalization, and characterization.

* Department, Faculty of Sciences and Natural Resources, University of Malaysia Sabah

Y. Umul Hanim *

Abstract- The successful in developing a straightforward preparative route in biomaterial composite fabrication with the great promising Umbilical Cord Matrix derived mesenchymal stem cells (MSCs) known as Wharton's Jelly also well known as fetal MSC. WJMSC has been created a reliable modernist and briskest innovative research method especially in manipulation of Graphene Oxide (GO) as a new biomaterials loaded with an ample advantages in regard of synthetic materials that applicable for a replacement or regenerating a living system. Bio-interface in mesenchymal stem cell certainly brighten the hope of all stripes scientist researchers in aspiration of biomaterial application whilst can impact the benefit in the fields of biomedical, biotechnology, bioengineering, pharmacology and biosensors yet the finding remain a number of challenges in the fast-growing field. The purpose of this review, we enumerated the key aspect an earmark that will be focused on GO functionalization, and GO-WJMSC characterization, accordingly to GO-WJMSC application in enhanced the proliferation, and multipotency for a brighter future potential of GO-stem cells applicability.

Index Terms- Graphene Oxide, Graphene Oxide-based composite, substrate, Umbilical Cord Matrix Mesenchymal Stem Cells, Wharton's Jelly, functionalization, viability, multipotency, mesenchymal stem cell application.

1. Introduction

The studies of biomaterial substrates have been accepted without ample of questions for more than a generation (Baker, 2011) since it was discovered in the 1900s (Bergmann and Stumpf, 2013). The great field of biomaterial composite – stem cell technology in future advancement rely on the effective of robust designation throughout privilege materials composite integration. Therefore, the next horizon in the fast-growing field which in biomaterial composites exceptionally remain crucial challenges in elucidating the complexity understanding in advance modification of new germinated biomaterial namely graphene oxide (GO) loaded with complete advantages package particularly will enumerated in this review with the meaning of synthetic materials applicability in the promoting the stem cell behavior for instances in multipotency of stem cell (Lee et al., 2011: Nakajima et al., 2007:Luo et al., 2015:Kim et al., 2013) precisely serve as a transferable and implantable application evenly giving high impact of benefit in the fields of biomedical, biotechnology, bioengineering, pharmacology, biosensors (Park et al. 2015), energy conversion and storage (Li et al., 2015). Alternatively, exceptional innovative in GO composite attained a numerous attempting among scientist recently. Regardless in variety fields lured by the researchers, ideally the phase of clinical trial of biomaterials-human Mesenchymal Stem Cells (hMSCs) still infancy stage even the discovery of GO .

MSCs therapy relies on a larger-scale of derivative sources in the various preparations to biomedical application as revealed in tissue engineering, drug delivery, gene delivery and bioimaging (Rahul et al. 2016). Study from Zhao et al. 2015 dictated that MSCs can be isolated from umbilical cord, namely Wharton's Jelly up to a point of other sources as surely from bone marrow, placenta, fat, amniotic fluid (Kim et al. 2015) and also from urinary membrane (Bharadwaj et al. 2011) and enable to comply in 3-dimentional, 3D innovation without restricted regardless stem cell behaviors. WJMSCs attained a workable source basically in tissue engineering approach because it readily isolated, expended, cell direct injection and transplanted either intra-peritoneum or intravenous (Neuss et al., 2008; Nakia et al., 2010:Gu et al., 2014: Zhou et al., 2015).

Graphene Oxide (GO) biomaterial has been ideally proven as a novel material notably interest in the field of biomedical as for in vivo implantation application which vast in number of orthopedic therapy significantly (Garcia et al. 2016) in a position of its excellent amphiphilic compound properties specifically synthesis in a heavily oxidized by oxidation and exfoliation procedures of graphite (Liu et al. 2013; Rosa et al. 2016 ; Rahul et al. 2016; Park et al.2015; Lee et al. 2011:Gu et al., 2014), which is derived from the two-dimensional carbon from carboxyl group, epoxide and hydroxyl group, draw amazing effect to the physical, mechanical and biological properties of cell lines enable interaction by amphiphilic bonding (Shen et al. 2012:Li et al., 2015:Kim et al., 2013: Yang et al., 2009) illustrated in figure 1.

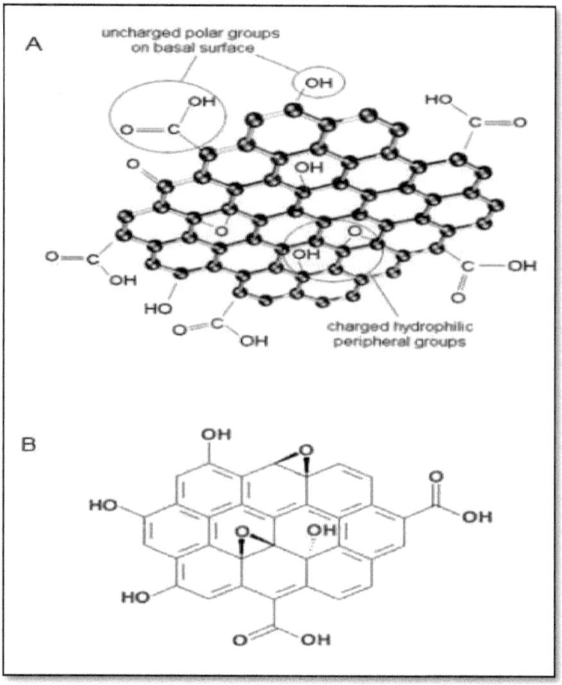

Fig. 1 Show the structure of GO structure, A) illustrated by Jastrzębska et al. 2012, B) Schematic illustrated by Pinto, et al., 2013.

As distinct study by Yang et al. 2016, stem cell engineering exploited using chemically exfoliations GO of single or multilayer atomic sheets as highly potential biomedical application due to physicochemical stability, high electrical conductivity, tunable amphiphilicity (Liu et al. 2013: Kim et al., 2015: Saravanan et al., 2017: Li et al., 2012: Dubey et al., 2015), mechanical strength (Yang et al. 2016: Kim et al., 2013) and good biocompatibility evenly allow control of cellular behaviors. The package of these unique properties of GO recently has driven the cellular behaviors for instance cell adhesion, proliferation, differentiation, polarization, and biocompatibility in MSCs on GO substrates or scaffold with regard to remain eliminated in severely declining stemness of adult and fetal derived human-MSCs during in vitro expansion on tissue culture polystyrene (TCPS) presently limits their therapeutic efficacy prior to cell transplantation (Lucas et a. 2016:Gu et al., 2014) by manipulation of natural and synthetic biomaterials innovation, therefore, various attempting in incorporating interface of GO-based biomaterial composite modification have been crystalized to envision the novel finding of efficacy in stem cell behavior. Undoubtedly, the purpose of this review herein to be enumerated the key aspect issues that will be projected on GO functionalization, GO-WJMSC characterization, and GO-WJMSC application in enhanced the MSCs proliferation, and multipotency also future potential of GO-stem cells towards applicability.

2. Graphene Oxide

Since early 2004, GO has effectively modified the process synthesis from the current fabrication including production using mechanical or thermal exfoliation, chemical exfoliation of graphite (Rosa et al., 2016: Sharma et al., 2016), chemical vapour deposition (CVP) and epitaxial growth (Gu et al., 2014). The oxidation of graphite to be graphite oxide is the threshold to fabricate the most effectively and friendly fabrication of GO by employed the mechanical exfoliation in water using sonication purposely to produce a single or few layers of GO whilst increase dramatically as highly desired by scientist due to a cheaper, simpler, more efficient, better quantity and quality compared to present method to be suitable financially for industrial and commercial larger purpose. GO also known as a flakes material that can be visible in yellowish suspension.

3. Graphene Oxide properties

Basically, the unique features of GO known as flakey material has aforementioned widely towards the several package properties enumerated as electrically insulator, physicochemical stability, mechanical strength, and biological compatible. Particularly, GO consist of three reactive oxygen functional group (ROFG) have lifter the properties of electrical insulator inclusively the epoxide, carboxyl and hydroxyl group. Electrically, GO plays a great insulating material due to the practicable of saturated sp3 orbital coordinate the missing carbon atom and negatively charged density species bound to carbon engendered GO as a super capacitor material (Li et al., 2015: Becerril et al., 2008). Schematic is illustrated in Figure 2.

The distinct physicochemical of GO inclusively listed as the easily dispersed due to attraction of hydrophilic, hydrophobic and amphiphilicity that encountered GO as a highly soluble and easy dispersible in water and organic solvent due to the ionizable edge –COOH affected by tunable of pH. That property has promises GO to be dispersed especially in physiological media making GO purposely apply in biomedical diligence (Cote et al., 2011: Kim et al., 2010). In addition, the unique properties of GO especially as a carbon based material lead to the elevation of hydrophilicity and large surface areas driven GO potentially utilize in drug delivery (Saravanan et al., 2017: Li et al., 2012: Dubey et al., 2015). Physically GO suspension can be soluble in water by express in dark brown to light yellowish in color regardless to any concentration.

Additionally, GO (Wang et al., 2011: Luo et al., 2015) assigned as a biocompatibility materials that coordinated the cellular behavior succinctly reported by Luo and colleagues, GO-incorporated Poly(lactic-co-glycolic acid) known as PLGA enhanced the cellular metabolism and osteogenic differentiation importantly as a protein adsorption through the capacity of surface interaction. Due to these unique properties, GO is inevitable to be a favor candidate to be delved in the chemical modification in order to create a better adaptable for a lot of application by functionalization depending on desired application. Accordingly to the study by Wang and group stated that GO in low (0.1mg/ml) and mild dose 0.25mg/ml) is activate the cell growth and consider as a high biocompatibility (Wang et al., 2011).

4. Wharton's Jelly Derived Mesenchymal Stem Cell

Mesenchymal stem cells (MSCs) parentage is from a layer of mesoderm has been proved of the high capacity for self-renewal and great in multi-lineage differentiation into adipocytes, osteoblasts, chondrocytes and smooth muscle cell (Zhou et al., 2015: Pittenger et al., 1999: Wang et al.,2004). MSCs primitively derived from bone marrow also well-known as an adult stem cell discovered comparable to fetal stem cell evenly fetal stem cell efficacy possible to encounter more beneficial than adult stem cell. MSC has been circulated in the meeting agreement of International Society Cellular Therapy (ICST) by Dominici et al. (2006) which as MSC has three minimal criteria to be defined as truly MSC population, firstly MSC must be plastic-adherent when maintained in standard culture conditions. Secondly must capable to express CD105, CD73 and CD90, and lack expression of CD45, CD34, CD14 or CD11b, CD79 alpha or CD19 and HLA-DR surface molecules. Lastly MSC enable differentiate to osteoblasts, adipocytes and chondrocytes in vitro. Exceptionally, fetal stem cell has studied possibly derived from multiple tissues and organ for instance umbilical cord (Martin-Pendon et al., 2008), placenta (In 't Anker et al., 2004), amniotic fluid

(Tsai et al., 2004), and Umbilical cord blood (Jang et al., 2006). A nifty study of human Wharton's Jelly derived umbilical cord matric has brighten the field of MSCs with regard to the special advantage of readily available, free ethically, easily harvested, non-invasively pathogenic and enrichment resources (Zhou et al., 2015). Basically, hWJMSCs are concern to exhibit outstanding advantages over other MSC resources especially for cell-based tissue engineering, cell therapy and regenerative medicine (Pittenger et Al., 1999: Zhao et al., 2015). Successfully stem cell therapy has requested billions of hWJMSC in their application however there is an inevitable factor which may lead to the loss of multipotent ability and rely to cell senescence in cell-based therapy or regenerative medicine. Thus, the multitude effort should be designed for the development of a novel finding and technology that to enhance the yields of hWJMSC while maintaining the proliferation potential and multi-potency of the cell.

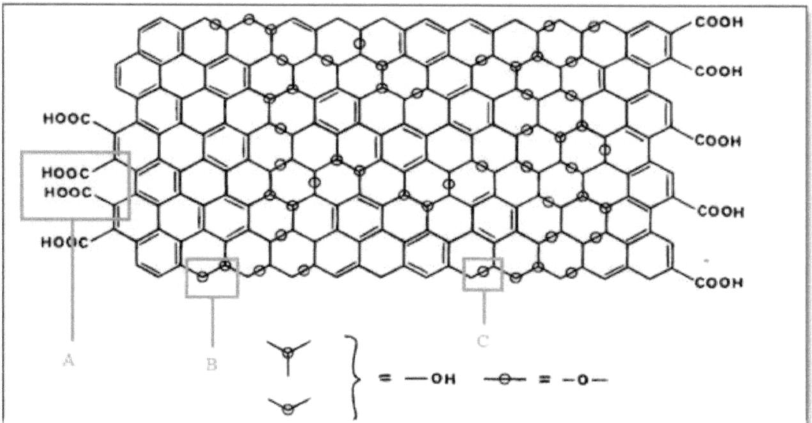

Fig. 2 Fictitious the most preferable structure of GO with three reactive oxygen functional groups. A: Pairwise carboxyl groups, B: Hydroxyl groups, C: Epoxy bridges. Source by Lerf et al., 1998.

Fig. 2 Schematic of reactive oxygen functional group (ROFG) of GO-based composite adopted by Lerf et al., 1998.

5. GO-based composite functionalization

Along the lines of GO modified functionalization method, strong emphasize in the superior properties in completely package is enumerated for instance as a unique functional group applicable including epoxide, carboxyl and hydroxyl group may lead to serve a larger surface area, mechanical stability, electrostatic interaction capacity and optical properties which promise its great characteristic among other materials ideally expended commitment on GO application as scaffold either substrate including in vast field of biotechnology, biomedical, biosensor and stretched a head upon in energy platform also in lithium batteries and as a super capacitors (Becerril et al. 2008; Li et al. 2015: Sharma et al., 2016).

The most common GO synthesis is manufactured using first principal by Hummers method implied the oxidation of graphite treated with potassium permanganate and sulfuric acid (Gu et al., 2013: Hummers et al., 1958) nonetheless GO has been considered as one of a low-cost material manufacturing which attract the wide world scientist to give full effort on GO investigation (Dubey et al., 2015: Singh etal., 2009: Sharma et al., 2016).

Generally, there are several common on a solid support substrate have been demonstrated in tissue engineering for instances including glass, polymers, gold or silicon, and thin coatings such as poly-L-lysine, agarose or polyacrylamide (Ghaemi et al., 2013) to a slight variation in emphasize the substrate functionalization through dip or spin coating, plasma treatment, surface grafting, self-assembled monolayer (SAM) formation and micro fluidics approaches.

To address the huge amount of literature in composites functionalization innovatively, herein as to generally summarize that there are enumerate of the key factors as to contemplate the efficacy of composites for instance synchronization integrated factor exhibited the advancement effect of cell phenotypes, materials compositions manifestation (Nancy et al. 2016; Singh et al., 2009). Next, for the second factor is the extended homogeneity compositional material interaction whereas reliability control the cell lines behavior as a 3D scaffold, substrate, or implantation devices (Fisher et al. 2010; Nakajima et al. 2006). For the third factor is the sustainability of biocompatible in versatile technique in the cell-material interaction which detail manifested by Neuss and colleagues in 2008 analyze the variation parameter stem cell material combination for tissue engineering. Views of the above points, seem is loaded with the meaning of materials composite, combination and hybrid material purposely.

Deeply entail on the GO composite owing a huge meaningful of the advantage potential along the materials combination for instances such nanoparticles, nanocomposite (Mahmoudi et al., 2016; Ji. et al., 2016; Zuo et al, 2013), nanofibers (Mahmoudi et al. 2016), and polymer scaffold (Kim et al., 2015). Regardless the vast approaches of GO composite innovation namely construct modification in Extracellular Matrix (ECM) component, administrate of growth factors (Nakajima et al., 2007; Lee et al., 2016), surface and functionalize modification and included the compilation biomolecule features interaction (Ghaemi et al., 2013), in fact there a still remain an ample challenges in promoting and complying the approaches. Ideally the combination should provide chemical stability, physical matching properties, adherent support, and mechanical strength of surface topography and surface modification capacity (Sabine et al., 2008).

The fascinating properties of GO mostly correspondence to its various functional group applicability for instance epoxide, carboxyl and hydroxyl group have prone to amphiphilicity interaction and excellent in transmittance of thin film thickness evenly in bio-composite. Interestingly, that sense, GO ultimately can be functionalized to enhance bioactivity of the substrate or commonly to be used as a surface coating on biomaterials. This is a great promising of surface chemistry modification which serves as a high anticipative tool in regulation of biological responses meantime sustain the stem cell behavior including cell adhesion, viability, proliferation and cell spreading determinate biocompatibility of the certain substrate (Kang et al., 2017: Baker, 2011: Ning et al., 2016). In that reason, we elicited the GO functionalization accordingly to the most common application of stem cell-biomaterial interface illustrated as cell biological application, bio-device, biosensor, bio-environment and so forth.

5.1 Cell biological application

In vitro culture

In general, many of this application frequently based on the unique surface activity between GO-cell cultures which is reflected in physicochemical properties to some extent is amphiphilicity enumerated as hydrophilicity, hydrophobicity, and covalent or non-covalent interaction and to some extent via natural biomaterial mimicry extracellular matrix (ECM). The biological system surface functionalization of GO is therefore pivotal to be construed the solubility and biocompatibility of three ROFG purposely for in vitro cell culture (Makharza et al., 2013: Lin et al., 2011). Some study has prompted to modification on hydrophobicity limitation. There are several methods have delved to alter the interaction of hydrophobicity may include the physically protein adsorption or in short by natural ECM biomaterials interface and covalently modification (Ning et al., 2016: Wu et al., 2006: Vu et al., 2014). To address the modification via physical protein adsorption, hydrophobic surface need to be altered in order to reduce nonspecifically adsorbs serum protein (albumin) from cell which hindered the affinity of cell adhesion. Certainly, Fibronectin (FN), laminin and collagen are the conventional physical protein modification. Basically, chemical modification enables to stimulate the robust interaction between protein and substrate surface (Ning et al., 2016: Kuddannaya et al., 2013). The specific cross-linker possibly to use such as (3-aminopropyl) triehoxy silane (APTES) and cross-linker glutaraldehyde (GA) to modified surface enable cell adhere easily (Lin et al., 2011) especially for cell adhesion and viability of MSCs providing the effective surface modification which reduce the hydrophobicity.

Li et al in 2015 have addressed the huge expression of excellent candidate in incorporating with other materials has essentially fabricated by the functionalized graphene with oxygen-containing chemical group created highly potential as a platform that serve as biocompatible, transferable and implantable device for innovative stem cell culture, additionally there are large surface area for the immobilization culture system where as the substrate enable to modified especially in the application of biomolecules, fluorescent, and drug carrier (Lee et al., 2011). By Kim et al., 2013-explained ITO, functionalized with micrograph via spectro-electrochemical tool C-G.

Atomically, epoxide (−C=C−), hydroxyl group (C=O-OH), and carboxyl group (C (O) OH) (Rahul et al., 2016; Rosa et al., 2016; Lee et al., 2011) have notably ascertained a greater interactions with protein membrane adsorption through covalent, electrostatic, and hydrogen bonding whereas pivotal in accelerate the stem cell behavior especially in adhesion differentiation in osteogenicity activity.

Deep in studied by Mokkapati et al. (2016) was presented that human Adipose Stem Cells (hASCs) positively growth and well proliferate on a new class of membrane integrated with GO and Boron (NaB). The study of integration GO-NaB towards the method of alternatively consume of ethanol (EtOH) for dispersion GO flakes on top of a glass substrate uniformly revealed highly in biocompatibility in relation to squeezed the cost for cultivation of cell and tissue study.

Owing to the fact that dispersion, Liu et al. (2013) in the review also belief the important of functionalization oxygen-containing hydrophilic groups assist high dispersion in water solvent. Correspondence study with Garcia et al. (2016) stated that the successful uniform evidence of 100% monolayer GO flakes revealed by analysis of both is a powerful surface analytical technique by atomic force microscopy (AFM) and scanning electron microscopy (SEM) in outstanding of oxidation and exfoliation GO.

Dramatically revealed from Erickson and the team, have discovered that the specific atomic features have contributed three major regions illustrated as hole, graphitic, and high contrast of oxygen functionalization density which recapitulating the various functionalization applicable (Erickson et al., 2010: Chen et al., 2012). In addition , the actual figure of structurally GO regardless beneficial to stem cell was elucidated that excellent physiochemical properties that can be seen as ultra-large surface areas, abundant hydrophilic groups and outstanding water solubility give rise in amazing effect of manipulation in cell adhesion, cell-cell interaction, and cell spreading which exhibit from the exceptional GO characteristics specifically in mechanical stiffness, nanotopography and large absorption energy capacity (Park et al. 2015).

Functionalization group of GO alternatively discovered by Rahul et al. (2016) illustrated the nucleophilic addition–elimination reaction whereas to insist the uniform dispersion of nanoparticles in resulting of increasing the adhesive factor for cultivation and developed implantation field. To address the amazing result from this method, high density polyethylene hybrid to polyethylene grafted graphene oxide (HDPE/ PE-g-GO) indicated the highest adsorption of protein on the substrate surface reinforcing obviously with GO substratum compared to unmodified HDPE.

Alternatively, overall scientists are straggling an effort to envisioned the multiple novel approach in regulating the acceleration of stem cell behavior which as according to La and the team also followed by Ji and the groups have been first successful in demonstrated the first time of combining layer-by-layer (LBL) technology with the biomimetic mineralization method (La et al., 2014; Ji et al., 2016).

Behind another issue especially in biocompatibility herein stretched in reviewed by Pinto et al. (2013) epitomize the amazing property of GO as excellent in biocompatibility by highlighted stripes issues for instance are in vitro studies of biocompatibility in mammalian cell and bacterial cells as well as Hemocompatibility in remarkable analysis of GO and composites GO materials in order to distinguish the effect of GO employed in biomedical field of implantation and intravenous application.

Moreover, putting in an expedient work, the promising composites studied by Rahul et al. 2016 successful to develop the combination of nano-composites reinforcement by GO with HDPE/PE-g-GO indicate the highly protein adsorption which established well mechanism interaction of electrostatic and hydrophobic interaction with the substrate upon increasing the interfacial adhesion and proliferation cell culture between matrix and filled. The modification technique by reinforced with GO, ideally enhance the growth numbers of hMSCs in analysis of cytocompatibility assessment using osteoblast and MSCs of cell attachment result the high number of mitochondria in the cells. The great quality of cell morphology also disparity between the control of unmodified with GO by the amazing discovery indicates that the great effort and high commitments in bio-composites candidate giving wider chances in regenerative biomedical loaded to orthopedic application evenly projected.

Myoblast differentiation

Study by Chaudhuri et al., 2015 expelled the unique promising of properties GO physically rougher also has enhanced not only osteoblast hence demonstrated that GO composite Poly-caprolactone (GO-PCL) significantly elevated the proliferation and differentiate in to human skeletal muscle cells (hSkMCs).

In the same perspective, Ku and Park, 2012 has demonstrated that GO on amine modified glass substrate was cleaned by Piranha solution and immersed then modified in 3-APTES in anhydrous toluence solution has shown a great capability to proliferate mouse myoblast C2C12. GO-modified glass substrate has delineated succinctly to express specific genes instantly MyoD, myogenin, Troponin T and myosin heavy chain (MHC). The study has demonstrated the ability of GO-modified glass substrate enable to stimulate myogenic differentiation giving the future lightens in tissue engineering application.

Neuronal differentiation

For the purpose of surface modification of GO for neuronal stem cell, Yang et al. (2016) was grounded explain the other way of functionalization technique on substrate before coating GO which as self-assembled monolayer of positively charged 3-aminopropyltriethoxysilane (APTES) employed on the surface by immersed into APTES solution dissolve in distilled water (DI). Contrary, the elaboration relied on consumed of silicone (SiO_2) as a substrate for GO fabrication in a reason of typically template applied for stripes of cell cultures due to the power of biocompatibility. Alternating of outstanding idea on replenish the modification by surface hierarchical topographies cue for instance GO patterned substrate significantly exhibited an expression result in neuronal differentiation using neuronal markers (TUJ1 and MAP2) and promoting the focal adhesion development in human neural stem cells (hNSCs). Customarily, due to the electrical interaction upon the group of carboxyl whereas negatively charged with self-assembled monolayer of positively charged on the substrate passionately innovation the line-patterned samples. In the line to confirm the electrophysiological functionalities, Yang and the group were being performed a whole-cell patch clamp analysis. The neuronal cells derived from hNSCs adhered on GPS revealed the voltage-dependent ionic currents and action potentials which are the electrophysiological characteristics treated using sodium channel antagonist tetrodotoxin (TTX) about 5-10min purposely to examine whether the currents and spikes were Na^+ channel specific Yang et al., 2016.

Dental

In contrary, discovered by Lee et al. (2015) ideally report that reduced GO (rGO) as applied in dental and orthopedic field mediated spontaneously stimulate osteogenic differentiation of MSCs which as composite with hydroxiapatite (HAp). The result revealed that the composites obviously stimulate the multipotency of stem cell culture and showed the superior bioactivity and biocompatibility with the average of particle sizes of HAp and rGO were 1270 ± 476 nm and 438 ± 180 nm, respectively. The study commence the that rGO-HaP has high potential in dental and orthopedic field.

Owing the unique properties of GO, Akhavan et al., 2012 have rendered the study of GO alone interface to dental pulp stem cell. There would be the first study to projected the specific concentration of 1.5mg/ml GO solution has added to the coverslip which enable to allowed the attachment, proliferation and presented the positive expression for all genes tested such as PAX-9, COL 1 and DMP-1. Furthermore, this pioneer study on dental pulp stem cell reckoned that the rougher surface of GO has attracted the stem cell attachment rather than normal coverslip.

The report of studied by Rosa and colleague has expressed that GO alone obtained by chemical exfoliation of graphite using the modified Hummer's method enable to elevate the growth of dental pulp stem cell (DPSC). The rougher GO surface capable DPSC to express the gene expression of Paired box-9 (PAX-9), Collagen type I (COL 1), Dentin matrix acidic phosphoprotein 1 (DMP-1) and Homo sapiens dentin sialophosphoprotein (DSPP) (Rosa et al., 2016).

Ref.	Method modification	Functionalization/composite	Cellular types	Approach	Application
Rosa et al., 2016	GO Modified by Hummer's Method	NA	Dental pulp stem cell (DPSC)	GO concentration of 1.5mg/mL. Odontoblastic differentiation, substrate for DPSC.	Dental
Mokkapati et al., 2016	GO flakes on glass substrate	boron (NaB)	human adipose stem cells (hASC)	Concentrations of NaB and GO are 0.02% and 1/20 of stock (0.024%) respectively.	In vitro culture
Li, M. et al., 2016	Ti substrate	Silk fibroin Hydroxyapatite Graphene oxide (SF/GO/HA)	human osteosarcoma MG63 cells	Exploition electrophoretic nanotechnology with SF being used as both a charging additive and a dispersion agent.	In vitro culture
Rahul et al., 2016	GO sheets	Polyethylene (PE) modified high density polyethylene (HDPE) based composites.	osteoblast cell line (MC3T3) human mesenchymal	Dispersion by nucleophilic addition–elimination reaction	In vitro culture

			stem cells (hMSC)		
Yang et al., 2016	GO-sample surface electrophysiology (via electrostatic interaction)	Fabrication of GO-based patterned substrate	Human neural stem cells (hNSCs).	Samples were functionalized with a self-assembled monolayer of positively charged 3-aminopropyltriethoxysilane (APTES) andsubstrates were coated with fibronectin (FN).	Neuronal differentiation
Lee et al., 2015	Reduced graphene oxide (rGO)-coated hydroxyapatite (HAp) composites	(HAp)	hMSCs	Synthesis the modification of Reduced GO.	Dental and orthopedic
Garcia et al. 2016	GO coated cover slip substrate	-	Mouse embryonic Stem (ES) cells.	Obtaining excellent distribution of 100% monolayer GO flakes by oxidation and exfoliation procedure.	In vitro culture
Park et al., 2015	GO flakes added in culture media	GO suspension	Human Bone Marrow MSCs	Synthesis GO flakes using a modified Hummers method. (1µg/mL-10µg/mL).	Cardiac repair
La et al., 2014	Ti substrate coating with 0.05 % (w/v) GO	Layer by layer (Ji et al., 2016) modified on Ti with GO and growth factor inducer	Bone marrow MSCs	Positively charged graphene oxide (GO-NH3+) was prepared via the 1-[3-(dimethylamino) propyl]-3-ethylcarbodiimide methyliodide and ethylenediamine (EDC) mediated amine exchange reaction.	In vitro culture
Akhavan,2012	GO sonicated suspension	rGONPs	Human umbilical cord (hUCMSC)	Cyto-toxicity, geno-toxicity, cell viability	In vitro culture
Tatavarty et., 2014	GO	calcium phosphate nanocomposite	Mesenchymal stem cell (MSC)	Osteogenesis	Orthopedic
Yang et la., 2013	GO/modified Hummers method	-	Embryonic stem cell (ESC)	No significant different of toxicity GO (1um/ml-100um/ml). Dopamine neuron differentiation.	In vitro culture
Singh et al., 2013	GO	EPD technique for GOPC (polymeric isocyanate crosslinked with hydroxyl functional acrylic, PIHA).	-	Protection from oxidation and corrosion by providing hydrophobic coating	In vitro culture
Zhou etal., 2016	Go-Ti substrate	Immersing the APTES-treated sodium titanate substrates with an elestrostatic interaction.	Periodontal ligament stem cell (PDLSCs)	Bioactivity of human PDLSCs on GO-Ti substrate compared to Sodium-Titanate (Na-Ti) substrate.	Dental
Lee et al., 2011	GO film prepared by Langmuir Blodgett (LB)	GO-dip coating on PDMS substrate	Human BMMSCs	GO was dispersed in water methanol (1:5) solution at a concentration of 1 mg/mL. 100µL of the GO solution was deposited on the water-filled LB. The film was compressed to a surface pressure value of 20 mN/m.	Orthopedic and tissue engineering

*rGONPs: reduced Graphene Oxide Nanoplatelets, EPD: Electrophoretic deposition, GOPC: Graphene oxide-polymer composite,PDMS: Polydimethylsiloxane.

Table 1 Method functionalized modification in specific application in stem cell culture.

5.2 Biomedical device application

In this account, the discussion will evenly addressed on the idea of controllable physically, chemically, and biologically properties might potentially elevating upon directing hMSCs especially toward the various endeavor vary in biomedical device purpose (Ghaemi et al. 2013; Yang et al., 2016). The high focusing in the interrelationship between the processing, structure designing, and performance compatibility cell-material interface underlie the basis rationality for most biomedical device application issues.

Commonly, the first highlighted in the key factor of medical application can be enumerated as matrices or substrate controllable degradation features as essential ability in facilitate the morphological and cell-substrate interaction especially in proper cell interface signaling which currently shortcoming in sustainable the mechanical stability *in vivo* degradation upon developing of tissue engineering (O'Brien, 2011; Council regulation, 2008: Gu, et al., 2014).

The second key factor describes the potential bio-adhesive matrices or substrates importantly to allow cell interface to biomaterials adhere to the surface materials in fact either by using linker molecules and functionalization groups in evidence to analyze the features of cell adherent effectiveness including the specific differentiation effect. Customarily, ample variation of drug delivering system applied the great candidate of GO with MSCs cell lines due to high loaded capacity of drugs and molecules in fact using simple modification procedures of GO (Liu et al. 2013). Obviously the unique amphiphilic structure owned by GO has driven many researchers over the world in practically the great innovation in medical development either in improvement in biocompatibility and avoid totally the cytotoxicity interruption as broadly explore the potential of GO application in biomedical engineering and biotechnology subsequently (Pinto et al. 2013; Kim et al. 2015: Kotsuki et al., 2014).

Basically the innovation in orthopedic therapy focused the fundamental of similar function and physiology in natural bone which discussed by Rahul et al. (2016) that bone in nature compose of the composite interaction with collagen, polymer matrix, hydroxyapatite as inorganic reinforcement and essentially the artificial innovation of composite in biomedical should significantly match with the natural bone filled. In the same position, Nayak et al. (2011) also agreed that the present application of GO in biomedical elucidated the well incorporation cell-material interface primarily in implantation or injured tissue as its free in side effect to the physiological condition of the microenvironment. Due to the biocompatible attractive, GO also well establish in composite with Hydroxyapatite (HAp) benefit to biomedical application especially in dental and bone implantation device as GO and its derivative promote the osteogenic differentiation administrated the mineralization of calcium and phosphate absorption (Lee et al. 2015; Jurczyk et al; Anderson et al. 2016; Zhou et al, 2016; Lozano et al., 2014: Gu et al., 2014).

Advances studied by Park et al. (2015) demonstrated that GO-flakes in sight permitted the amazing benefit to the treatment of heart disease as ability of GO flakes to prevent the adverse adherent in cell-signaling cascade in the response of reactive oxygen species (ROS) which generate the poor survival MSCs intravenous application in myocardial infarction (MI) by figure out with *in vitro* culture using the similarity of adverse effect of agar as a blocking agent coating on the plate in order to idealize the same function in reality of heart condition with the present of anoikis is a lesion site that inhibit the cell of MSCs adherent. Essentially, the study successively inspired the potential of GO-MSCs in cellular therapy implanted as taken a seed to treat the ischemic disease or tissue regeneration.

In relation, Mo and colleagues in 2016 studied the advance functionalization GO with Dopamine and subsequent bioglass (BG) in order to express the outstanding effect in cell adhesive and osteogenic differention as to commit invented of curative medicine to bone neither replacement nor reconstruction. Doubtless, Yang and colleagues also successively harnessing the potential of GO in biomedical by using the modification on synergistic surface topographical in enhancement and improvement neuronal lineage commitment for the effective culture platform to stretch the treating neurodegenerative diseases and neuronal disorders (Yang et al., 2016).

Furthermore, advanced features study by La et al. in 2014 indicated the GO-Ti substrate facilitated by Bone Morphogenetic Protien-2 (BMP-2) implanted MSCs successful demonstrated to get squeezed on tissue regeneration and organ transplantation among bone defected. In view of future invasive of MSCs expansion, Zhao et al. in 2015 have been demonstrated the large-scaling of Wharton's Jelly (WJ-MSCs) disparity study along 3-dimensional features of cells envisioned to treat skin reconstruction such skin injury.

Interestingly, Mokkapati and co-workers strongly discovered the integration GO as a new class membranes composites with Boron (NaB) subsequently elevated the cell viability of human Adipose Stem Cells (hASCs) upon 3D-GO in the future. In view of intuition idea on combination, Ming Li et al. (2016) nonetheless evaluate the robust combination of mechanical property by demonstrated the idea of combination between silk fibroin (SF), HAp and GO on the Titanium (Ti) substrate importantly to reveal the novel coating for orthopedic application. SF/GO/HA on Ti substrate established as first verified composites by explore the strength of electrophoresis nanotechnology which combine the benefit of SF as confer a charging additive and dispersion agent.

A part from that, the innovation of biomedical device has been delved of the fabrication of GO-based single-bacterium, label-free Deoxyribonucleic Acid (DNA) sensor, bacterial DNA/protein and polyelectrolyte chemical transistors whereas successively studied by Mohanty and berry (2008).

5.3 Biosensor application

Biosensor application germinated accordingly to the unique GO properties especially in feasibility functionalization, highly sensitive in electrical properties and easily disperses in aqueous medium (Chung et al., 2012: He et al., 2010: Sharma et al., 2016). Basically, fundamental of biosensor encompassed of two parts for instance receptor is a molecule which interacts with an analyte and second is a transducer as a sensing platform to convert the chemical information into a signal (Sharma et al., 2016: Nurunnabi et al., 2015). Essentially, the important of biosensor application is being referred to the detection of biomolecules namely nucleic acids (NAs), protein, and growth factors that have been demonstrated successively by using appropriately functionalized graphene derivatives with NAs, aptamer and carbohydrate in order to monitor target specific changes electrically.

Demonstrated study by Lu et al., 2010 has shown the unique ability of GO enable to detect a multiple Deoxyribonucleic acid (DNA) manage to bind the dye-labeled via non-covalent binding. The presence of a target molecule thereby has altered the conformation of DNA by disturbing the GO DNA interaction. The result elucidate that the dye-labeled DNA is released from the surface of GO and restore the fluorescence of dye. Theoretically has revealed that both DNA and GO are negatively charged the fluorescence of carboxyfluorescein (FAM)-tagged DNA which was rapidly quenched by GO. Lu and team have delineated that the advantage of this biosensor has involves the articulate syntheses of GO with large planar surface renders the feasibility to analyze multiple molecular targets in the same solution, rapidly and sensitively at low-cost effectively. Regarding the adsorption of DNA on GO surface, critically to elaborate the important to improve the existing design of biosensor according to several interactions including π π stacking, hydrophobic interaction and hydrogen bonding have been attributed to the binding process.

5.4 Bio-environment and energy application

The promises of GO shown by nanomaterials and advance coatings technique prone to increase further expectation especially in the larger industries of marine, pipeline, aerospace, automobiles and construction. Currently there are wider applications of GO nanocomposite coating, hydrophobic coating, and also organic-inorganic hybrids have been envisioned to improve the survival of materials under stringent or severe environment condition. Studied by Singh and group has witnessed the robust oxidation and corrosion resistant coatings of graphene oxide-polymer composite (GOPC) containing GO and polymeric isocyanate crosslinked with hydroxyl functional acrylic adhesive as polymer matrix is used to prepare endure oxidation and corrosion has enabled the possibility of designing environmental friendly anti-corrosion, inexpensive, simple cathodic constant voltage electrophoretic deposition (EPD), high density coating, uniform thickness and devoid the cracks (Singh et al., 2013). The significant of the study reflect the excellent oxidation and corrosion resistance characteristics under severe chlorine ion normally in marine engineering materials which require very high salt resistance.

Study of possibility of GO to remove the aniline compound from aqueous solution has been succinctly discovered by Fakhri (2013). The unique properties of GO have prone the wide application of GO as an alternative adsorbent for the well-known of chemical compound broadly applied in various manufacture of dyestuffs, rubbers, pesticides, plastics and paints. The adverse effect of the compound has affected the human health and severely harms the environment. The experimental analysis via thermodynamic parameter which including free energy change, enthalpy change, and entrophy change witnessed that the removal of aniline from GO was endothermic and spontaneously. Hence, the study construed that GO capable to be used as an efficient adsorbent material for the adsorption of aniline from aqueous solution.

In the same line, studied by Li and Zhang, 2015 also has reckoned the outstanding properties of GO prone to be a promising nanomaterial for energy and environmental application. Functionalized grapheme with oxygen-containing chemical groups, has encountered presently special elucidating superior properties such as large surface area, mechanical stability, tunable electrical and optical properties. Exceptionally, GO surface functional group containing of hydroxyl, epoxy and carboxyl rendered GO an excellent candidate in coordinating with other materials or molecules intentionally hold a versatile applications of energy storage/conversion and environment protection including hydrogen storage materials, photo-catalyst for water splitting, removal of air pollutants and water purifications for instance in various lithium batteries and super-capacitors.

5.5 Drug carrier application

Yang et al., 2009, have addressed the issue of GO application as a drug carrier is reflected from the high capacity loading by GO properties. Owing to the unique structure encompasses of one-atom thickness and two-dimensional plane render with a large adjustable specific surface area for the deposited immobilization rely the excellent candidate of a large number of substances including a wide range of metals, biomolecules, fluorescent molecules, drug and so forth for targeted drug delivery. They had been concluded that GO-Fe3O4 hybrid with a simple, effective and scalable method to chemically deposit Fe3O4 nanoparticles onto GO. This hybrid can be loaded with anti-cancer drug doxorubicin hydrochloride (DXR) with a high loading capacity up to 1.08 mg mg^{-1}. The super-paramagnetic property of the hybrid has revealed the flexible dispersible and enable to congregate under acidic and reversibly under basic condition due to pH-triggered controllable behavior makes this material a promising candidate for controlled targeted drug delivery witnessed in biomedicine, biomaterial and bio-diagnostics.

Furthermore, there is various attempted have been fictional regarding to the great witnessed of GO functionalization to precisely obtain the specific application in drug delivery. Sun and his team have presented the finding of nano-graphene oxide (NGO) synthesis specialty to cellular imaging and drug delivery. The NGO visualized being photo-luminescent in the visible and infrared regions. GO functional group of epoxide has been activated with chloroacetic acid under strongly basic conditions in order to convert hydroxyl groups to carboxylic acid (-COOH) moieties (Sun et al., 2008) whereas has elevated water solubility to be available subsequently pegylation. To this end, NGO-Polyethylene glycol (PEG) has succinctly conjugated covalently with a B-cell specific antibody Rituxan (anti-CD20) to NGO-PEG(NGO-PEG-Rituxan) in order to selective recognize the binding to B-cell lymphoma cell and prone to be a potential product for cellular biology and drug carrier.

6. GO-WJMSC characterization analysis

To date, in order to delve the GO-based composite in more depth, various analysis and techniques have been employed for the probing of its efficacy towards WJMSCs or any stem cell. In the sight of fundamental of GO composites, there are considered as great promising biomaterial applicable as a scaffold in tissues engineering, substrates for stem cells cultivation and lured implantation device components (Lee et al. 2015). To regard the ample features of GO composite, herein we fascinatingly will elaborate the wise discovery from the variety previous research papers in disparity of GO-WJMSC or GO-MSC characteristics and application.

GO-based material escorted as the basic conceptualization in its characterization of myriad analysis for instance studied by Sun et al. (2008), has projected the thickness of GO is ~1.0nm for the single-layered structure using atomic force microscope (AFM). In addition, analyzed using Infrared (IR) spectroscopy has fictitious the existence of –OH (~3400cm^{-1}), C=O (1715cm^{-1}), and C=C (1580cm^{-1}) accordingly to functional group in GO (Hummers et al., 1958: Mahmoudi and Simchi 2016: Hontoria-Lucas et al., 1995).

Despite the curiosity interlinking with exciting focusing in the trendy of stem cell interface with promising biomaterial variation, there are eventually underneath of challenges in this fast-growing cell-biomaterials signaling. Stem cell hold precious complexity requirement due to their unique ability in self-renewal and differentiation especially in Mesenchymal stem cell (MSCs) ideally as a cell source of current tissue engineering innovative incorporated exceptionally regarding to ease in readily isolated, expanded capability and transplanted applicability (Anderson et al. 2016;Ghaemi et al., 2013). Hence the minimal requirement of MSCs verification is considerable natively adhere in tissue culture polystyrene (TCPS), yet the cell condition cannot fully elucidated in extended of sustainability of culture condition especially the purity and multipotency of stemness. Mostly myriad of biomaterial, GO especially draw fascinating interaction with living cells either *in vivo* or *in vitro* in view of trendy culture in research field towards the outstanding in soft membrane sheet with high in plane stiffness whereas served as a biocompatibility, and implant ability for stem cells (Hofmeister et al., 2015; Lee et al. 2011). The district ability of GO as a great loading capacity and higher electrostatic interaction presence in oxygenated groups functionalization, delineated by successively play a major role in facilitating cell adhesion whereas resulted higher adsorption for serum proteins induced cell proliferation and osteoblast differentiation exclusively.

6.1 Morphology

Analysis microscope

Analysis using transmission electron microscopy (TEM) by Erickson et al. (2010) has breakthrough the discovery which identify the illustration of specific atomic scale features of the GO material compared to pristine graphene. There are comprised three major features presently as holes indicated in blue, graphitic region indicated in yellow and highly contrast disordered regions indicated in red. Holes regions are released by the process of aggressive oxidation and sheet exfoliation and second regions are graphitic. Notwithstanding the high contrast oxidized regions of the basal plane exhibited a continuous network throughout the GO sheet. In the high contrast regions elucidated the present of higher density and greater numbers of oxygen atoms will be mobilized by the electron beam which may renders the oxygen functionalities with each carbon. Fig. 3 and 4.

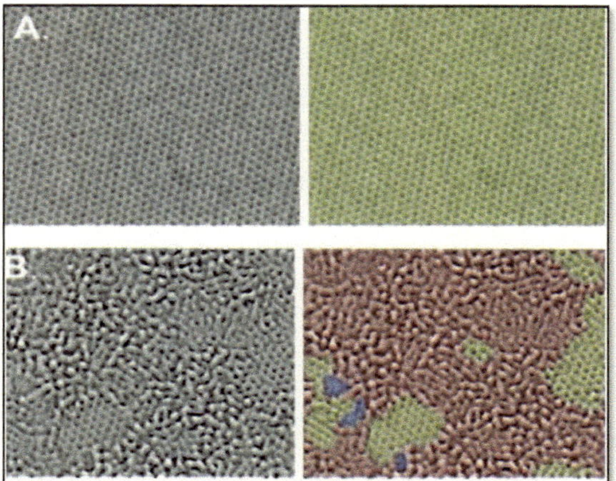

(A) Single sheet of graphene. On the right, graphitic area indicated in yellow. (B) Single sheet of GO. On the right, graphitic areas in yellow, and high contrast, disordered regions, indicating oxygen functionalities, in red. Adopted from Erickson et al., 2010.

Figure 3 depicted the illustration of TEM analysis of GO sheet indicating the different region of ROFG adopted by Erickson et al., 2010.

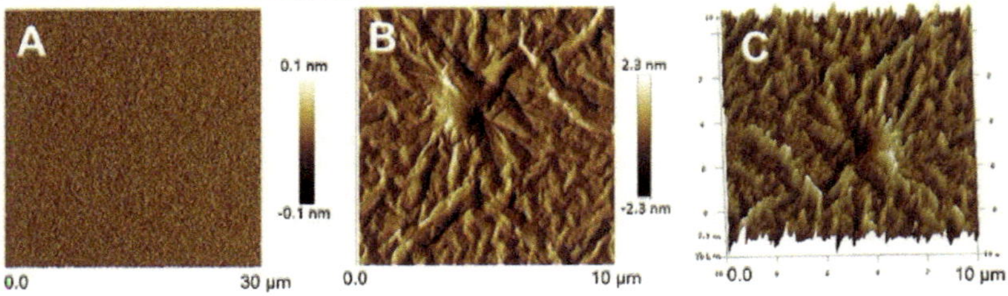

Atomic force microscopy (AFM) (A) on glass surface, (B) GO and (C) GO provided a rougher surface.

Figure 4 Morphology of rougher surface GO compared to glass surface. Source by Rosa et al., 2016.

6.2 Viability

Generally, Mokkapati and co-workers have strongly proved the GO-based composite with Boron (NaB) enable subsequently elevated the cell viability of human Adipose Stem Cells (hASCs) due to latest class membranes integration upon GO composites as a highly biocompatible material and greatly cost-effective for cell culture studies (Mokkapati et al., 2016). The process of dispersion the GO flakes on the top of glass substrate was assisted by ethanol (EtOH) purposely for uniformity. The study was the novel finding for the effort of cell attachment and cell proliferation among the population of MSC as to provide the development of 3D-GO scaffold system for tissue engineering.

Studied by Elkhenany and team, was discovered the outstanding finding of GO coated plates compared to standard polystyrene-coated tissue culture plates succinctly support cell adherent and proliferation of caprine-MSCs (c-MSCs). Exceptionally, cell seeded on to oxidized graphene films undergo osteogenic differentiation without specific growth factor engagement. Cell viability on the GO substrate revealed >95% at each passage when the cells were harvested at 70%confluency deliberately to maintain their optimal density for retaining the growth and proliferation condition (Elkhanany et al., 2014).

6.3 Phenotypically and purity

Actual figures of MSCs are presented as multipotent undergo cell differentiation into multiple cell types which as acquisition of cell type-specific morphology followed by phenotypic expression. There are limited in circumstances importantly describes the condition to evolve the actual features of MSCs whereas has been declared by the Mesenchymal and Tissue Stem Cell Committee of the International Society for Cellular Therapy (ISCT) that minimal criteria to fulfill the definition of human MSC as firstly, MSC must be flawless on plastic-adherent whilst maintained in standard culture conditions. Secondly, MSC must enable phenotypically express by markers of CD105, CD73 and CD90, and lack expression of CD45, CD34, CD14 or CD11b, CD79 alpha or CD19 and HLA-DR on surface molecules. Lastly, MSC must be capable differentiated to osteoblasts (bone), adipocytes (fat) and chondrocytes (cartilage) in vitro (Pittenger et al. 1999; Gabrielis, 2012). Collectively, they are able to be derived from multiple tissues in particularly from bone marrow, bone tissue, fat, umbilical cord and some research reported can be isolated from dental tissues.

Putting together the fundamental of MSCs devoted as therapeutic agents that generate fascinate excitement in recent research trend due to their high plasticity, great self-renewal and wide-lineage potential with attractive immunosuppressive properties (Bernard and Sangeeta, 2004; Joss *et al.*, 2015; Alexander, 2014).

The notion that leads to the effect of cell –material interface, Rahul and co-workers successful report that the strongly uniform dispersion of nanoparticles towards the interfacial adhesion leads to the increasing of cell viability for hMSCs at day 7 cultured on modified with GO reinforced substrate by indicator of increment almost 30% - 40% of yield strength of 20Mpa and elastic modulus of 600Mpa respectively. In a similar position, hMSCs appear in well cell adherent to substrate modified GO due to the successful effort on modification on functional group accordingly to the interaction of carboxylic groups (C=O), epoxy ring (-C –C), and hydroxyl group (O-H).

Great discovery upon GO stratum conceded by Mo and group interestingly dictate that dopamine functionalized with GO followed by bioglass fabrication (DGO/BG) using sol-gel method draw stimulation of rate bone marrow mesenchymal stem cells (rBMSCs) adhesion, cell spreading, and osteoinduction *in vitro*. Within 2 days cultivation of rBMSCs, cell on 1 wt. % and 5wt. % significantly showed more spreading and expending area towards fast confluence percentage (Mo et al. 2016). The result studied by Elkhenany et al., 2014 parallel expressed the caprine adult MSCs on GO substrates enable to positively expressed for CD29 (delta Ct 0.19±0.001) meanwhile CD90 (delta Ct 2.9±0.0015) and expressed negatively for CD4 and CD45 apparently prove that these cells are non hematopoietic merely only population of mesenchymal lineage.

On the basis study of GO coated titanium (GO-Ti) substrate compare to sodium titanate (Na-Ti) substrate, Zhou et al. 2016 indicated that human periodontal ligament stem cells (PDLSCs) impress inhibited significantly higher proliferation rate and promote protein expression related to the bone sialoprotein (BSP), runt related transcription factor 2 (Runx2) and osteocalcin (OCN) comparatively with Na-Ti substrate. The relation finding report by Li et al. (2016) in the same indicator as a good adhesive of GO composites by enhance the alkaline activity (ALP) of human osteosarcoma MG63 cell line on SF/GO/HA nano-composite biomaterials promising a precious orthopedic implantation therapy.

In spite of large investigation draw on the in vitro application, Park et al. (2015) presented both in vitro and in vivo application to analyze the notable benefit of GO flakes in curative medicine of MI by demonstrated the MSCs-GO implantation successfully decrease the infarction size of left ventricle by the assisted of GO-flakes adherent competitively than the lesion site of anoikis. The finding reveals the free-toxicity of GO to human organ applicability. Table 2 is demonstrated the GO-based composite interface to MSCs accelerate the stemness.

6.4 Multi-potency

Generally, wider concern about several stage of MSC differentiation division had been studied by Huang et al., 2007: Hoemann et al., 2009 stated that MSCs ensured to be expressed a new bone formation whenever cultured in osteogenic media which are the cells responsible for laying down the matrix and mineral during new bone formation *in vivo*. There are three stages of MSCs osteogenic differentiation *in vitro*. The first stage consists of day one to four where a peak in the viability number of cells is seen. Secondly is an early cell differentiation from day 5 to 14, which is characterized by the transcription and protein expression of alkaline phosphatase (ALP) (Huang et al., 2007). The final stage from days 14 to 28 indicate the results in a high expression of osteocalcin and osteopontin, followed by calcium and phosphate deposition (Hoemann et al., 2009; Huang et al., 2007).

In convergence, Lee and the group successively report that reduced GO (rGO) mediated spontaneously stimulate differentiation of MSCs which as composite with hydroxiapatite (HAp). The result revealed that the GO-based biocomposites obviously stimulate the multipotency of MSC culture and showed ultimately superior bioactivity and biocompatibility for some extent employed in various field such as in dental (Rosa et al., 2016) and orthopedic field (Lee et al., 2015: Tatavarty et al., 2014).

Koroleva and co-workers emphasized about the hybrid material organic-inorganic based towards the scaffold with human MSCs. The bioactive porous 3D polymer ceramic scaffold, zirconium-silicone (Zr-Si) indicated the successful fabrication herein enhance the kinetic growth and differentiated into bone or dental importantly in implantation device due to their 'cell-culture friendly' (Koroleva et al. 2015).

Primitively aforementioned, MSCs population owing a unique muti-lineage accoutered to differentiate into several types of cell inclusively osteoblasts, adipocytes, chondrocytes, and myoblasts (Kalbacova et al., 2010: Ku and Park, 2013: Zeddou et al., 2010: Kalaszyzska and Ferdyn, 2015). The aggregation of both tremendous benefits prompted the WJMSC osteo-differentiation capability on substrate or scafflold materials by surface modification. Fascinatingly, proved by Mo and groups, 1 wt. % DGO/BG substrate engaged the stimulation of higher expression osteogenic genes in immunofluorescence staining of runt-related transcription factor 2 (RUNX2), and osteopontin (OPN) compare to other samples after cultivation of 7 days respectively (Mo et al. 2016). However, contrary study revealed by Nayak and group in 2011 showed that the substrate graphene alone provides a promising biocompatibility scaffold that does not hamper the proliferation of hMSCs and revealed the remarkable differentiation of specific bone cells in presence under the influence of Bone Marrow Protein-2 (BMP-2) (Nayak et al., 2011). (Fig.5)

In remind of the structurally unique of GO characters, herein create a sense of high potentially as a biocompatible, transferable, implantation capable and adhesion dispersible especially in MSCs and DPSCs in the relation studied by Lee et al. (2011) and Rahul et al. in 2016 dictated that growing of MSCs accelerated by GO substrate as an artificial scaffold and ability to differentiate toward the osteogenic lineage by elevation of dexamethasone in parallel increase the mineral deposition extended the differentiation osteogenicity by expression in Alizarin Red S staining in the same point to adipogenic lineage due to the electrostatic expressed in higher affinity for insulin mainly as a mediator to fatty acid synthesis successively indicate higher accumulation of lipid for adipogenic differentiation (Dubey et al., 2015: Akhavan et al., 2013).

In the sight of GO enhanced in differentiation, Rosa and co-workers took an advance investigation on benefit of roughness GO substrate to DPSCs cell lines which successful investigating the attachment and cell proliferation within 3-5 days cultivation assessed. Moreover, gene expression successfully interpreted on GO substrate compared to glass solely as control.

Tentatively studied by Zancanela et al. (2016) on the valuable optimization of GO sheet to enhance the better differentiation of osteoblast in rat bone marrow (rBM) have given the result up to 170% of cell viability after 21 days of incubation involved concentration of 25ug/mL GO reveal outstanding osteoblast viability grown on plastic surface compare to Ti disc as a control.

The insight studied by La and co-workers in 2014, exceptionally the advancement of implanted Ti-GO substrate modified by using Bone Morphogenetic Protein-2 (BMP-2) demonstrated the enhancement in bone formation essentially support and improve the osteoblast differentiation accordingly. Contrary, demonstrated by Yang et al. in 2013 stated that GO potentially in exhibited the dopamine neural differentiation of Embryonic Stem Cells (ESCs) and envisioned in cell transplantation therapy. The study about GO incorporated to undifferentiated stem cell get a critical concern whereas discovered the valuable tool to characterize the distinguished differentiation of stem cell (McMurray et al., 2011: Kim et al., 2013). Surface-enhanced Raman spectroscopy (SERs) is being known as a powerful chemical/biological analysis of reagent-free, rapid and non-destructive tool while persist stem cell viability and suitable for *in vivo* studies. This may lead the easier tool to identify the undifferentiated stem cell generally compose of polyunsaturated membranes and unsaturated molecules such as chromosomes, proteins and lipids that are rich in C-C bonds due to aromatic structures. Table 2.

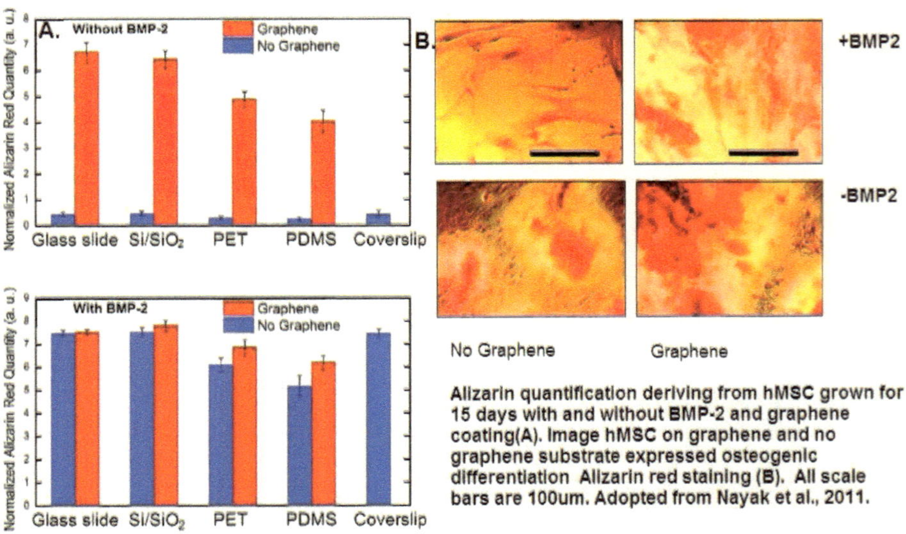

Alizarin quantification deriving from hMSC grown for 15 days with and without BMP-2 and graphene coating(A). Image hMSC on graphene and no graphene substrate expressed osteogenic differentiation Alizarin red staining (B). All scale bars are 100um. Adopted from Nayak et al., 2011.

Figure 5 Illustrated the qualification and quantification of alizarin staining on several substrates with and without BMP-2 (A), Graphene expressed slightly darker in alizarin staining with BMP-2 (B) sources by Nayak etal., 2011.

Ref.	Material composite	Cellular types	Characteristics	Outcome
Mokkapati et al., 2016	boron (NaB)	human adipose stem cells (hASC)	Proliferation	Better cell viability on NaB integrated GO membranes.
Rahul, et al., 2016	Polyethylene (PE) grafted GO (PE-g-GO)	osteoblast cell line (MC3T3) human mesenchymal stem cells (hMSC)	Protein adsorption Proliferation Multipotent	High in protein adsorption by modified HDPE/PE-g-GO 3%. Highest cell growth of HDPE/PE-g-GO 3% and supported osteoblast growth.
Saravanan et al., 2017	Chitosan and Gelatin	mouseMSC	Protein absorption Biomineralization Cytotoxic assessment Cell viability Cell morphology	GO/CS/Gn improveporous structure. 0.25%GO/CS/Gn enhanced protein absorption. Cytocompatiblewith no signs of toxicity. 0.25%GO/CS/Gn promotes osteoblast differentiation.
Park et al., 2015	GO flakes added to culture media.	human mesenchymal stem cells (hMSC)	Cytotoxicity	Indicated negative cytotoxicity of GO flakes at concentration below 20*u*g/mL. Generate GO flakes-MSCs cellular implanted for Myocardial Infarction (MI).
Lee et al., 2011	GO film	human mesenchymal stem cells (hMSC)	Cell proliferation Osteogenic differentiation Adipogenic differentiation	Higher density of cells on GO than on the PDMS substrate. GO film adsorbed protein up to 28% on day 1. Higher exhibited of affinity for insulin by GO substrate as mediator for fatty acid synthesis enhances adipogenic differentiation.
La et al., 2014	Ti substrate with GO	Bone marrow MSCs	Osteogenic differentiation	Facilitated bone formation on Ti/GO/BMP-2 substrate implanted by Bone Morphogenic Protein-2 (BMP-2).
Ji et al., 2016	GO/Chitosan/Hydroxyapatite (GO/CTS-HA) nanocomposite film	Rat MSCs	Cell proliferation Immunostaining Cell morphology	HA presence to improve the biocompatibily of the film. Combination GO/CTS-HA enhanced the MSCs proliferation and good biocoating candidate.
Akhavan et al., 2012	rGONPs	Human umbilical cord blood MSCs	Proliferation	Cell viability significantly destructed by 1.0mg/mL rGONPs with average lateral dimensions (ALDs) of 11± 4 nm.
Luo et al., 2015	Electrospinning	Umbilical cord MSC	Hemocompatibility Viability Morphology RT-PCR	GO-doped PLGA enhanced hydrophilic performance. GO-doped PLGA eccelerate adhesion and proliferation compared to pure PLGA nanofiber.
Chaudhuri et al., 2015	Electrospun fibrous meshes of GO–PCL (poly-caprolactone)composite.	human cord blood derived mesenchymal stem cells (CB-hMSCs)	Immunostaining analysis Fluorescence-activated cell sorter (FACS) analysis Cells adhesion analysis Cell proliferation	A better biocompatibility of GO–PCL meshes might be associated with interconnectivity of fibrous meshes and enhanced σ and ε induced by GO, which might play an important role guiding cell adhe-sion, resulting in a higher proliferation and myotubes orientation
Elkhenany et al., 2014	0.1mg/ml Air-sprayed on specific size well to form oxidized grapheme (GO)	cBMMSCs	Cell adhesion Cell viability CFU assay Real-time PCR Trilineage differentiation	Cell onGO like phenotype on TCP. Trphan blue revealed >95% at each passage. cBMMSCs positive for CD29 (delta Ct 0.19± 0.001) and CD90 (delta Ct 2.9 ± 0.0015). Osteogenic, adipogenic and chondrogenic differentiation.

*PLGA:poly(lactic-co-glycolic acid),rGONPs: reduced Graphene Oxide Nano-pletelets, PCL: Polycaprolactone, cBMMSCs: caprine Bone Marrow derived MSCs,CFU: Colony-Forming unit, PCR: Polymerase chain reaction, TCP: Tissue culture plate.

Table 2 elaborate the characterization of umbilical cord and other MSC sources interface Go-based composites induced MSCs purity, viability, and multipotency.

7. GO-stem cell future potential applicability

To some extended observation, Mo et al. 2012 suggested that substrate composites of DGO/BG play as effective result in osteoinductivity and osteoconductivity which as recommended the further study regarding to their surface functionalization in order to proceed the successful result in future innovation. As deeply grounded by every researcher especially in biomaterial engineering respectively the novel technique still be widely to explore as discussed by Rosa et al. (2016), the biocomposites GO based preciously enhance certain bioactivity of GO for instance GO/HA will improve the corrosion resistant and reveal higher adhesion to titanium substrate or sheet while the composition of GO-BG enable to enhance the high rate of bone formation (Wan et al. 2011; Tatavarty et al. 2014; Wu et al. 2015).

8. Conclusion

As to explore strongly on modification GO-based composite modification applicability in stem cell cultivation neither in vivo nor ex vivo followed by advances innovation in bio-coating of 2-dimention, 2D and 3-dimention, 3D (Zhao et al., 2015) investigation play the crucial role to stretch the GO application in ample field for future initiation and innovation direct injection implanted or indirect transplantation in biomedical field and tissue engineering field as with full of concern on this review will enhance the future underlying regarding GO-based composite modification interface with myriad MSCs sources especially we have been focused on WJ-MSCs as to widespread the exception of particular properties in pioneer platform for future regenerative medicine and cell therapy.

Acknowledgment

This research was supported by fund from Trans-disciplinary Research Grant Scheme (TRGS) of Government Ministry of Malaysia Education.

References

[1] Lee, W.C., Lim, C.H.Y.X, Shi, H., Tang, L.A.L., Wang, Y., Lim, C.T., Loh, K.P. 2011. Origin of enhanced stem cell growth and differentiation on graphene and graphene oxide. American Chemical Society. Vol 5, No. 9: 7334-7341.

[2] Lee, J.H., Shin, Y.C., Jin, O.S., Kang, S.H., Hwang, Y.S., Park, J.C., Hong, S.W., Han, D.W. 2015. Reduced graphene oxide-coated hydroxyapatite composites stimulate spontaneous osteogenic differentiation of humen mesenchymal stem cells. Nanoscale. 7:11642-11651.

[3] Council regulation (EC). 2008. Setting up the joint undertaking for the implementation of the joint technology initiative on innovative medicines. No 73.

[4] Becerril, H.A, Mao, J., Stoltenberg, R.M., Bao, Z., Chen, Y. 2008. Evaluation of solution-processed reduced graphene oxide films as transparent conductors. ACS Nano 2: 463-470.

[5] Li, F., Jian, X., Zhao, J., Zhang, S. 2015. Graphene oxide: A promising nanomaterial for energy and environment applications. Nano Energy, 16: 488-515.

[6] Anderson, H.J., Sahoo, J.K., Uliin, R.V., Dalby, M.J. 2016. Mesenchymal stem cell fate: Applying biomaterials for control of stem cell behavior. Frontiers in Bioengineering and Biotechnology. Vol. 4: 38.

[7] Zhou, Q., Yang, P., Li, X., Liu, H., Ge, S. 2016. Bioactivity of periodontal ligament stem cells on sodium titanate coated with graphene oxide. *Sci. Rep.* 6: 19343.

[8] Pittenger, M.F., Mackay, A.M., beck, S.C., Jaiswal, R.K., Douglas, R., Mosca, J.D. et al. 1999. Multilineage potential of adult human mesenchymal stem cells. *Science* 284:143-147.

[9] Patrick, B., Clelia, C., Jens, K., Margit, S., Kathrin, H. and Edda, T. 2014. Stem cells on biomaterials for synthetic grafts to promote vascular healing. *Journal of Clinical Medicine* 3: 39-87.

[10] Blitterswijka, and Jan, D.B. 2011. An algorithm-based topographical biomaterials library to instruct cell fate. *PNAS October 4*, vol. 108, no. 40:16565–16570.

[11] Ghaemi, S.R., Harding, F.J., ,Delalat,B., Gronthos,B., Gronthos, S., Voelcker, N.H. 2013. Exploring the mesenchymal stem cell niche using high throughput screening. Biomaterials *34*:7601-7615.
[12] Josh, C., Mehdi, N., and David, A.B. 2015. Biomaterial approaches for stem cell-based myocardial tissue engineering. *Biomarker Insights* 10(S1):77–90.
[13] Fergal, J.O.B. 2011. Biomaterials and Scaffolds for tissue engineering. *Materialstoday*. Vol 14, Issues 3: 88-95.
[14] Alexander, M. R. 2014. Materials for the stem cell factories of the future. *Nature Materials*, 13 (6): 570-579.
[15] Bernard, O. P. and Sangeeta N. B. 2004. Tissue Engineering.*Pearson Prentice Hall Pearson Education*, Inc.
[16] Gabrielis, K. 2012. Surface markers distinguishing mesenchymal stem cells from fibroblasts. *ACTA MEDICA LITUANICA*.Vol 19. No.2:75-79.
[17] Shen, H., Zhang, L., Liu, M., Zhang, Z. 2012. Biomedical applications of graphene. *Theranostics*;2(3):283-294.
[18] Mo, X., Wei, Y., Zhang, X., Cai, Q., Shen, Y., Dai, X., Meng, S., Liu, X., Liu, Y., Hu, Z., Deng, X. 2016. Enhanced stem cell osteogenic differentiation by bioglass funtionalized graphene oxide substrates. Journal of Nanomaterials. Vol. 5613980.
[19] Rahul, U., Sharmistha, N., Nitu, B., Suryasarathi, B., Bikramjit, B. 2016. Modulation of protein adsorption and cell proliferation on polyethylene immobilized graphene oxide reinforced HDPE bionanocomposites. ACS Appl. Mater Interfaces. Vol. 8:11954-11968.
[20] Nayak, T.R, Andersen, H., Makam, V.S., Khaw, C., Bae, S., Xu, X., Ee, P.L., Ahn, J.H., Hong, B.H., Pastorin, G., Özyilmaz, B. 2011. Graphene for controlled and accelerated osteogenic differentiation of human mesenchymal stem cells. ACS Nano,5:4670–4678.
[21] Liu, J., Cui, L., Losic, D. 2013. Graphene and graphene oxide as new nanocarriers for drug delivery application. Acta Biomaterialia 9: 9243-9257.
[22] Li, M., Xiong, P., Mo, M., Cheng, Y., Zheng, Y. 2016. Electrophoretic-deposited novel ternary silk fibroin/graphene oxide/hydroxyapatite nanocomposite soatings on titanium substrate for orthopedic applications. Front. Mater. Sci.10:270.
[23] Wan, C., Frydrych, M., Chen, B. 2011. Strong and bioactive gelatin-graphene oxise nanocomposites. Soft matter. Vol. 7:6159.
[24] Tatavarty, R., Ding, H., Lu, G., Taylor, R.J., Bi, X. 2014. Synergistic acceleration in the osteogenesis of human mesenchymal stem cells by graphene oxide-calcium phosphate nanocomposites. Chem Commun. Vol. 50:8484.
[25] Wu, C., Xia, L., Han, P., Xu, M., Fang, B., Wang, J. 2015. Graphene-oxide-modified β-tricalcium phosphate bioceramics stimute in vitro and in vivo osteogenesis. Carbon. Vol. 93: 116-29.
[26] Zancanela, D.C., Simao, A.M.S., Francisco, C.G., Faria, A.N.D., Ramos, A.P., Goncalves, R.R., Matsubara, E.Y., Rosolen, J.M., Ciancaglini, P. 2016. Graphene oxide and titanium: synergistic effects on the biomineralization ability of osteoblast cultures. J. Mater Sci: Mater Med. 27:71.
[27] Kim, J., Kim, Y.R., Kim, Y., Lim, K.T., Seonwoo, H., Park, S., Cho, S.P., Hong, B.H., Choung, P.H., Chug, T.D., Choung, Y.H., Chung, J.H. 2013. Graphene-incorporated chitosan substrata for adhesion and differentiation of human mesenchymal stem cells. J.Mater Chem. B, 1: 933-938.
[28] Alegria, G.E., Iluit, M., Stefanska, M., Silva, C., Heeg, S., Kimber, S.J., Kuoskoff, V., Lacaud, G., Vijayaraghavan, A., Batta, K. 2016. Graphene oxide promotes embryonic stem cell differentiation to haematopoietic lineage. Scientific Reports. Vol. 6:25917.
[29] Yang, K., Lee, J., Lee, J.S., Kim, D., Chang, G.E., Seo, J., Cheong, E., Lee, T., Cho, S.W. 2016. Graphene oxide hierarchical patterns for the derivation of elestrophysiologically functional neuron-like cells from human neural stem cells. ACS Applied Materiala&interfaces. 17763-17774.
[30] Pinto, A. M., Goncalves, I.C., Magalhaes, F.D. 2013. Graphene-based materials biocompatibility:A review. Colloids and Surfaces B:Biointerfaces 111:188-202.
[31] Wan, G.L., Jin, M., Park, S., Yoon, H.H., Bhang, S.H., Char, K., Kim, B.S. 2014. Delivery of bone morphogenetic protein-2 and substance P using graphene oxide for bone regeneration. International Journal of Nanomedicine, 9:107-116.
[32] Hofmeister, L.H., Costa, L., Balikov, D.A., Crowder, S.W., Terekhov, A., Sung, H.J., Hofmeister, W.H. 2015. Patterned polymer matrix promotes stemness and cell-cell interaction of adult stem cells. *Journal Of Biological Engineering*, 9:18.
[33] La, W.G., Jin, M., Park, S., Yoon, H.H., Jeong, G.J., Bhang, S.H., Park, H., Char, K., Kim, B.S. 2014. Delivery of bone morphogenetic protein-2 and substance P using graphene oxide for bone regeneration. International Journal of Nanomedicine, 9:107-116.
[34] Neuss, S., Apel, C., Buttler, P., Denecke, B., Dhanasingh, A., Ding, X., Grafahrend, D., Groger, A., Hemmric, K., Herr, A., Dechent, W.J., Mastitskaya, S., Bouza, A.P., Rosewick, S., Salber, J., Woltje, M., Zenke, M. 2008. Assessment of stem cell/biomaterial combination for stem cell-based tissue engineering. Biomaterials 29:302-313.
[35] Nakajima, M., Ishimuro, T., Kato, K., Ko, I.K., Hirata, I., Arima, Y., Iwata, H. 2007. Combinatorial protein display for the cell-based screening of biomaterials that direct neural stem cell differentiation. Biomaterials, 1048-1060.
[36] Park, J., Kim, B., Han, J., Oh, J., Park, S., Ryu, S., Jung, S., Shin, J.Y., Lee, B.S., Hong, B.H., Choi, D., Kim, B.S. 2015. Graphene oxide flakes as a cellular adhesive: Prevention of reactive oxygen species mediated death of implanted cells for cardiac repair. American Chemical Society. Vol. 9. No.5:4987-4999.
[37] Zhao, G., Liu, F., Lan, S., Li, P., Li, W.,Kou, J., Qi, X., Fan, R., Hao, D., Wu, C., Bai, T., Li, Y., Liu, J.Y. 2015. Large-scale expansion of Wharton's jelly-derived mesenchymal stem cells on gelatin microbeads, with retention of self-renewal and multipotency characteristics and capacity for enhancing skin wound healing. Stem Cell Research & Therapy, 6:38.
[38] Koroleva, A., Deiwick, A., Chichkov, B. 2015. Osteogenic differentiation of human mesenchymal stem cells in 3-D Zr-Si organic-inorganic scaffolds produced by two-photon poymerization technique. PLoS One, 10(2): e0118164.
[39] Mahmoudi, N., Simchi, A. 2016. On the biological performance of graphene oxide-modified chitosan/polyvinyl pyrrolidone nanocomposite membranes: *In vitro* and *in vivo* effects of graphene oxide. Materials Science and Engineering C 70:121-131.
[40] Ji, M., Li, H., Guo, H., Xie, A., Wang, S., Huang, F., Li, S, Shen, Y., He, J. 2016. A novel porous aspirin-loaded $(GO/CTS\text{-}HA)_n$ nanocomposite films: Synthesis and multifunction for bone tissue engineering. Carbohydrate Polymers 153:124-132.

[41] Lozano, F.J.R., Bernal, D.G., Cervantes, S.A., Roca, M.A.R., Alguer6, M.C., Atucha, N.M., García, A.A., Moraleda, J.M., Cenis, J.L. 2014. Effects of composite films of silk fibroin and graphene oxide on the proliferation, cell viability and mesenchymal phenotype of periodontal ligament stem cells. J Mater Sci: Mater Med, 25:2731-2741.
[42] O'Brien F. 2011. Biomaterials & scaffolds for tissue engineering. Materials Today. Vol. 14, Issue 3:88-95.
[43] Zuo, P.P., Feng, H.F., Xu, Z.Z., Zhang, L.F., Zhang, Y.L., Xia, W., Zhang, W.Q. 2013. Fabrication of biocompatible and mechanically reinforced graphene oxide-chitosan nanocomposite films. Chemistry Central Journal, 7:39.
[44] Dubey, N., Bentini, R., Islam, I., Cao, T., Neto, A.H.C., Rosa, V. 2015. Graphene: A versatile carbon-based material for bone tissue engineering. Stem Cells International. Vol. 15, 3.
[45] Singh, V.K., Patra, M.K., Manoth, M., Gowd, G.S., Vadera, S.R., Kumar, N. 2009. In situ synthesis of graphene oxide and its composites with iron oxide. New Carbon Materials. Vol. 24, Issue 2, 24(2):147-152.
[46] Yang, D., Li, T., Xu, M., Gao, F., Yang, J., Yang, Z., Le, W. 2013. Graphene oxide promotes the differentiation of mouse embryonic stem cells to dopamine neurons. Nanomedicine, 9:2445-2455.
[47] Lee, N., Robinson, J., Lu, H. 2016. Biomimetic strategies for engineering composite tissue. Current Opinion in Biotechnology, 40;64-74.
[48] Kim,T.H., Lee,K.B., Choi,J.W. 2013. 3D graphene oxide-encapsulated gold nanoparticles to detect neural stem cell differentiation. Biomaterials 34:8660-8670.
[49] Gu,M., Liu,Y., Chen, T.,Du,F.,Zhao,X.,Xiong,C.,Zhou,Y. 2014. Is graphene a promising nano-material for promoting surface modificationof implants or scaffold materials in bone tissue engineering?Tissue Eng., Part B, 20: 477-491.
[50] Kim,J., Cote,L.J., Yuan,W., Shull,K.R., Huang,J. 2010. Graphene oxide sheets at interfaces. American Chemical Society. 132:8180-8186.
[51] Cote,L.J., Kim,J., Tung,V.C., Luo,J., Kim,F., Huang,J. 2011. Graphene oxide as surfactant sheets. Pure Appl. Chem., Vol.83:95-110.
[52] Wang,K., Ruan,J.,Song,H., Wo, Y., Guo,S., Cui, D. 2011. Biocompatibility of graphene oxide. Nanoscale Res. Lett, 6:8.
[53] Luo,Y., Shen, H., Fang, Y., Huang, J., Zhang,M., Dai,J., Shi,X., Zhang,Z. 2015. Enhanced proliferation and osteogenic differentiation of mesenchymal stem cells on graphene oxide-incorporated electrospun poly(lactic-co-glycolic acid) nanofibrous mats. ACS Appl. Mater. Interfaces. 7:6331-6339.
[54] Baker,M. 2011. Stem cells in culture: defining the substrate. Nature Methods.Vol. 8:293-297.
[55] Lerf,A., He,H., Forster,M., Klinowski,J. 1998. Structure of graphite oxide revisited. J.Phys. Chem. B, 102: 4477-4482.
[56] Bergmann,C.P., Stumpf,A. 2013. Topic in mining, metallurgy and materialsengineering. *Dental Ceramics*. Springer-Verlag Berlin Heidelgerg. DOI: 10.1007/978-3-642-38224-6_2.
[57] Hummers,W.S., Jr., Offeman, R.E. 1958. Preparation of graphitic oxide. J Am Chem Soc. 80:1339.
[58] Ku,S.H., Park,C.B. 2013. Myoblast differentiation on graphene oxide. Biomaterials, 34:2017-2023.
[59] Kalaszczynska, I., Katarzyna, F. 2015. Wharton's jelly derived mesenchymal stem cells: Future of regenerative medicine?Recent findings and clinical signicance. BioMed Research International. Article 430847.
[60] Zeddou, M., Briquet, A., Relic, B., Josse, C., Malaise, M.G., Gothot, A., Lechanteur, C., Beguin, Yves. 2010. The umbilical cord matrix is a better source of mesenchymal stem cells (MSC) than the umbilical cord blood. Cell Biology International, 34; 693-701.
[61] Martin-Rendon, E., Sweeney, D., Lu, F., Girdlestone, J., Navarrete, C., Watt, S.M. 2008. 5-Azacytidine-treated human mesenchymal stem/progenitor cells derived from umbilical cord, cord blood and bone marrow do not generate cardiomyocytes in vitro at high frequencies. Vox Sang,95:137-148.
[62] In't Anker, P.S., Scherjon, S.A., Kleijburg-van der Keur,C., de Groot-Swings, G.M., Claas, F.H., Fibbe WE, et al. 2004. Isolation of mesenchymal stem cells of fetal or maternal origin from human placenta. Stem Cells, 22:1338-1345.
[63] Tsai, M.S., Lee, J.L., Chang, Y.J., Hwang, S.M.. 2004. Isolation of human multipotent mesenchymal stem cells from second-trimester amniotic fluid using a novel two-stage culture protocol. Hum Reprod., 19:1450-1456.
[64] Jang, Y.K., Jung, D.H., Jung, M.H., Kim, D.H., Yoo, K.H., Sung, K.W., et al. 2006. Mesenchymal stem cells feeder layer from human umbilical cord blood for ex vivo expanded growth and proliferation of hematopoietic progenitor cells. Ann Hematol. 85:212-225.
[65] Bharadwaj, S., Liu, G., Shi, Y., Markert, C., Andersson, K.E., Atala, A., et al. 2011. Characterizatio.n of urine-derived stem cells obtained from upper urinary tract for use in cell-based urological tissue engineering. Tissue Eng. Part A, 7:2123-2132.
[66] Wang JF, Wang LJ, Wu YF, Xiang Y, Xie CG, Jia BB, et al., 2004. Mesenchymal stem/progenitor cells in human umbilical cord blood as support for ex vivo expansion of CD34(+) hematopoietic stem cells and for chondrogenic differentiation. Haematologica, 89:837-844.
[67] Ning,R., Wang,F., Lin,L. 2016. Biomaterial-based microfluidics for cell culture and analysis. Trends in Analytical Chemistry, 80:255-265.
[68] Wu,D., Zhao,B., Dai,Z., Qin,J., Lin,B. 2006. Grafting epoxy-modified hydrophilic polymers onto poly(dimethylsiloxane) microfluidic chip to resist nonspecific protein adsorption. Lab Chip, 6:942-947.
[69] Vu,T.T.,Fouet,M.,Guec,A.M., Sudor,J. A new and easy surface functionalization technology for monitoring wettability in heterogeneous nano- and microfluidic device, sensor. 2014. Actuat.B.Chem., 196:64-70.
[70] Kuddannaya,S., Chuah,Y.J., Lee,M.H.A., Menon,N.V., Kang,F., Zhang,Y. 2013. Sulface chemical modification of Poly(dimethylsiloxane) for the enhanced adhesion and proliferation of mesenchymal stem cells. ACS App.Mater.Interfaces, 5:9777-9784.
[71] Yang,D., Velamakanni,A., Bozoklu,G. et al,. 2009. Chemical analysis of graphene oxide films after heat and chemical treatments by X-ray photoelectron and Micro-Raman spectroscopy. Carbon, 47:145–152.
[72] Makharza,S., Cirillo,G., Bachmatiuk,A., Ibrahim,I., Ioannides,N.,Trzebicka,B., Hampel,S., Rûmmeli,M.H. 2013. Graphene oxide-based drug delivery vehicles: functionalization, characterization,and cytotoxicity eluation. J. Nanopart Res., 15:2099-2125.
[73] Lin,Y., Jin,J., Song,M. 2011. Preparation and characterisation of covalent polymer functionalized graphene oxide. J Mater Chem. 21:3455.

[74] Mohanty,N., Berry,V. 2008. Graphene-based single-bacterium resolution biodevice and DNA transistor: interfacing graphene derivatives with nanoscale and microscale bio-components. Nano Lett. 8:4469–4476.

[75] Akhavan,O.,Ghaderi,E.,Shahsavar,M. 2013. Graphene nanogrids for selective and fast osteogenic differentiation of human mesenchymal stem cells. Carbon, 59:200-211.

[76] Kalbacova,M., Broz,A., Kong,J., Kalbac,M. 2010. Graphene substrates promote adherence of human osteoblasts and mesenchymal stromal cells. Carbon, 48:4323-4329.

[77] Akhavan,O., Ghaderi,E.,Akhavan,A. 2012. Size-dependent Genotoxicity of graphene nanoplatelets in human stem cells. Biomaterials, 33:8017-8025.

[78] Kotsuki,K., Tanaka,H., Obata,S., Stauss,S.,Saiki,K. 2014. The important of spinning speed in fabrication ofspin-coated organic thin film transistors: Film morphoogy and field effect mobility. Applied Physics Letter, 104:233306.

[79] Chen,D., Feng,H., Li,J. 2012. Graphene oxide: preparation, functionalization, and electrochemical applications. ACS. Chem.Rev. dx.doi.org /10.1021/cr300115g.

[80] Erickson,K., Erni,R., Lee,Z., Alem,N., Gannett,W., Zettl,A. 2010. Determination of the local chemical structure of graphene oxide and reduced graphene oxide. Adv. Mater., 22:4467-4472.

[81] He,Q., Sudibya, H. G., Yin, Z., Wu, S., Li, H., Boey, F., Huang, W., Chen, P., Zhang, H. 2010. Centimeter-long and large-scale micropatterns of reduced graphene oxide films: fabrication and sensing applications. ACS Nano, 4:3201–3208.

[82] Chung,C.,Kim,Y.K., Shin,D., Ryoo,S.R., Hong,H.B., Min,D.H. 2012. Biomedical applications of graphene and graphene oxide. Accounts of chemical research. 10.1021/ar300159f.

[83] Yang, X., Zhang,X., Ma,Y., Huang,Y., Wang,Y., Chen,Y. Superparamegnetic graphene oxide-Fe_3O_4 nanoparticles hybrid for controlled targeted drug carriers. *J.Mater.Chem.*, 19:2710-2714.

[84] Hontoria-Lucas,C., Lopez-Peinado,A.J., Lopez-Gonzalez,J.D.D., Rojas-Cervantes,M.L., Martin-Aranda,R.M. 1995. Study of oxygen-containing groups in a series of graphite oxides: Physical and chemical characterization. Carbon, 33:1585-1592.

[85] Sun,X., Liu,Z., Weisher,K., Robinson,J.T., Goodwin,A., Zaric,S., Dai,H. 2008. Nano-grapene oxide for cellular imaging and drug delivery. *Nano Res.*, 3:203-212.

[86] Saravanan,S., Anjali,C., Vairamani,M., Sastry,T.P., Subramanian,K.S., Selvamurugan,N. 2017. Scaffolds containing chitosan, gelatin and graphene oxide for bone tissue regeneration *in vitro* and *in vivo*. Int. J. Biol. Macromol. http://dx.doi.org/10.1016/j.ijbiomac.2017.01.034.

[87] Li,X., Liu,H., Niu,X., Yu,B., Fan,Y., Feng,Q., Cui,F.Z., Watari,F. 2012. The use of carbon nan otubes to induce osteogenic differentiation of human adipose-derived MSCs *in vitro* and ectopic bone formation in vivo. Biomaterials 33: 4818-4827.

[88] Dubey,N.,Bentini,R., Islam,I., Cao,T., Neto,A.H., Rosa,V. 2015. Graphene : a versatile carbon-based materials for bone tissue engineering. Stem Cells Int. 1.

[89] Rosa,V., Xie,H., Dubey,N., Madanagopal,T.T, Rajan,S.S., Morin,J.L.P., Islam,I., Neto,A.H.C. 2016. Graphene oxide-based substrate: physical and characterization, cytocompatibility and differentiation potential of dental pulp stem cells. Dental materials 32:1019-1025.

[90] Ryoo,S.R., Kim,Y.K., Kim,M.H., Min,D.H. 2010. Behaviors of NIH-3T3 fibro-blasts on graphene/carbon nanotubes: Proliferation, focal adhesion, and gene transfection studies. ACS Nano.4: 6587–6598.

[91] McMurray,R.J, Gadegaard,N., Tsimbouri,P.M. Burgess,K.V., McNamara,L.E.,Tare,R., Murawski,K., Kingham,E., Oreffo,R.O., Dalby,MJ. 2011. Nanoscale surfaces for the long term maintenance of mesenchymal atem cell phenotype and multipotency. Nat. Mater,10:637-644.

[92] Dominici.M., Le Blanc,K., Mueller,I., Slaper-Cortenbach,I., Marini,F., Krause,D., Deans, R., Keating,A., Prockop,D., Horwitz,E. 2006. Minimal criteria for defining multipotent mesenchymal stromal cells. The International Society for Cellular Therapy position statement. Cytotherapy, 8:315-317.

[93] Huang, Z., Nelson, E.R., Smith, R.L., Goodman,S.B. 2007. The sequential expression profiles of growth factors from osteoprogenitors [correction of osteroprogenitors] to osteoblasts *in vitro*. Tissue Eng. 13:2311-2320.

[94] Hoemann,C.D., El-Gabalawy,H., McKee,M.D. 2009. *In vitro* osteogenesis assays: influence of the primary cell source on alkaline phosphatase activity and mineralization. Pathol Biol 57:318-323.

[95] Yang,D., Li,T., Xu,M., Gao,F., Yang,J., Yang,Z., Le,W. 2013. Graphene oxidepromotes the differentiation of mouse embryonic stem cells to dopamine neurons. Nanomedicine (*Lond.*),9:2445-2455.

[96] Singh,B.P., Jena,B.K., Bhattacharjee,S., Besra,L. 2013. Development of oxidation and corrosion resistant hydrophobic graphene oxide-polymer composite coating on copper. Surface & Coating Technology 232:475-481.

[97] Farkhri,A. 2013. Adsorption characteristics of graphene oxide as a solid adsorbent for aniline removal from aquoes solutions: Kinetics, thermodynamics and mechanism studies. Journal ofSaudi Chemical Society. http://dx.doi.org/10.1016/j-jscs.2013.10.002.

[98] Li,F., Zhang,S. 2015. Graphene oxide: A promising nanomaterial forenergy and environmental applications. Nano Energy, 16:488-515.

[99] Elkhenany,H., Amelse,L., Lafont,A., Bourdo,S., Caldwell,M., Neilsen,N., Dervishi,E., Derek,O., Biris,A.S., Anderson,D., Dhar,M. 2014. Graphene supports in vitro proliferation and osteogenic differentiation of goat adult mesenchymal stem cells: potential for bone tissue engineering. Applied Toxicology, 35:367-374.

[100] Chaudhuri,B., Bhadra,D., Moron,L., Pramanik,K. 2015. Myoblast differentiation of human mesenchymal stem cells on graphene oxide and electrospun graphene oxide -polymer composite fibrous meshes: importance of graphene oxide conductivity and dielectric constant on their biocompatibility. Biofabrication 7 ,015009.doi:10.1088/1758-5090/7/1/015009.

[101] Ku,H.S., Park,C.B. 2013. Myoblast differentiation on graphene oxide. Biomaterials 34:2017-2023.

[102] Sharma,D., Kanchi,S., Sabela,M.I., Bisetty,K. 2015. Insight into the biosensing of graphene oxide: Present and futureprospects. Arabian Journal of Chemistry, 9:238-261.

[103]Nurunnabi,M.D., Parvez,K., Nafiujjaman,M.D., Revuri,V., Khan,H.A., Feng,X., Lee,Y.K. 2015. Bio-application of graphene oxide derivatives: drug/genedelivery, imaging, polymeric modification, toxicology, therapeutics and challenges. RSC Adv. 5(52): 42141-42161.

[104]Lu,C.H., Zhu,C.L., Li,J., Liu,J.J., Chen,X., Yang,H.H. 2010. Using graphene to protect DNA from cleavage during cellular delivery. Chem. Commun, 46(18):3116-3118.